SCIENCE
Magic
in the bathroom

in the bathroom

Written and illustrated by

Richard Robinson

OXFORD

These books are dedicated to two households: firstly the one I grew up in, where Dad, Mum, Anne, John and Philip were for ever dazzling their gullible youngest with tricks like the ones here.
Secondly the household I got for myself later, with Morgan and Georgia providing a new and demanding young audience for these same tricks.

OXFORD
UNIVERSITY PRESS

Great Clarendon Street, Oxford OX2 6DP

Oxford University Press is a department of the University of Oxford.
It furthers the University's objective of excellence in research, scholarship, and education by publishing worldwide in

Oxford New York

Auckland Bangkok Buenos Aires Cape Town Chennai
Dar es Salaam Delhi Hong Kong Istanbul Karachi Kolkata
Kuala Lumpur Madrid Melbourne Mexico City Mumbai Nairobi
São Paulo Shanghai Taipei Tokyo Toronto

Oxford is a registered trade mark of Oxford University Press
in the UK and in certain other countries

First published in 1999
This edition 2003

British Library Cataloguing in Publication Data available

ISBN 0-19-911153-7

1 3 5 7 9 10 8 6 4 2

Printed in UK

CONTENTS

Introduction

All the tricks in this book are self-working – that means you don't have to be a great magician to do them. The 'magic' will be done by Nature.

Magic and science have a lot in common.

Both magic and science can produce wonderful effects that leave us gob-smacked. Audiences always want to work out how a magic illusion works. Scientists try equally hard to understand Nature's tricks,

Magicians use a lot of misdirection – getting the audience to look in one direction while the trick is being done in another; Nature often seems to be doing the same. For thousands of years we thought that the Sun travelled across the sky above us; now we know that the Sun stays put and we do the travelling, so the Sun only seems to move. That's Nature misdirecting us.

A magician's audiences will often say 'I know how that's done!', when in fact they've got it completely wrong. Scientists often make the same mistake. Two thousand years ago, the Greek philosopher Aristotle had some pretty wrong ideas. For instance, he thought that apples fell from trees because they wanted to. Aristotle's ideas seem crazy nowadays, but for 1500 years everyone thought he was the top banana!

Aristotle's mistakes have been corrected now, but some of the magic in these books still can't be explained even by the best of today's scientists; that makes it doubly magical.

As soon as a scientist finds that an experiment has gone wrong, he starts again, looking and testing and guessing until he gets it right. As you practice these tricks you'll find that they sometimes go wrong, but with a little practice you'll get them right. Soon your tricks will seem as magical to your audience as Nature seems to scientists.

Good luck.

Richard Robinson

CHAPTER ONE

WEIRD WATER

Water is weird – perhaps you don't realise how strange it is . . .

It can behave like a glue.

It can become completely invisible.

Seconds later it can be completely opaque – you can't see through it.

Seconds later it can act like a mirror.

Then it can climb out of its container!

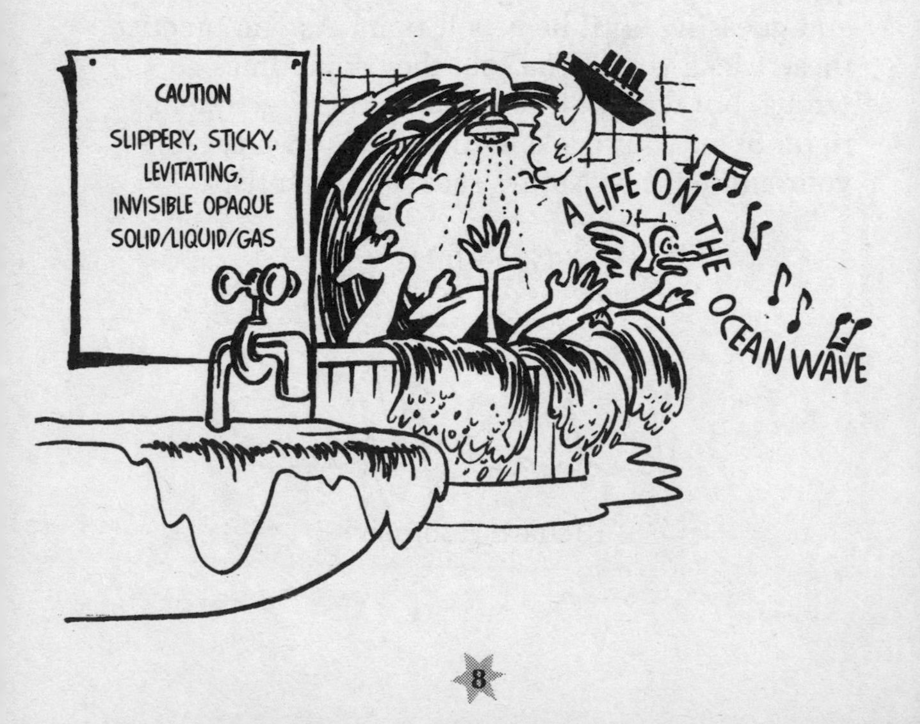

All of these magic properties can be used to make stunning illusions. And that's before we mix soap with it and do even more incredible tricks.

Then there's the magic of sound; why do things sound different in the bathroom to anywhere else in the house? Then there's the magic of heat and cold; why do you shiver when you get out of a hot bath? Where do goosebumps come from? Why do ttttttteeth chchchchchchatter? You'll find out later in the book.

Parents are going to tell you that the only magic they want to see in the bathroom is your skin turning skin-coloured again at the end of the day, and when are you going to finish in there, so they can rub seaweed juice and paint (see toothpaste, page 45) all over their teeth, or remove the grease from their skin and hair, and then put it all back (see pages 34 and 46) etc.

But first, let's do a little magic . . .

FLOATING NEEDLE

Water has a skin! Next time you pass a pond look closely at it. You will probably see insects walking on the surface of the water. Their feet bend the 'skin', but don't break through.

TADA-A-A-A!

We can use the pond skaters' trick to make some Big Magic.

THE EFFECT
The magician makes a needle float on water.

YOU NEED
- Bowl of water
- Needle
- Tissue or toilet paper

TO PERFORM
Challenge your audience to float a needle in a bowl of water. When they can't manage it they'll mutter something about it being impossible, because metal's heavier than water. That is when you make your announcement; you have extraordinary power over water, and if you tell it to let metal float, then that's what it must do!

Fix the bowl of water with a steely stare for five seconds; concentrate on getting it to obey you.

Now make a little raft of tissue paper, put the needle on it and float it on the water.

When the tissue is soaked it will sink, leaving the needle sitting there on the surface. You can see the dent in the water's skin.

Push the pin and it will break through the skin and sink, like a pin should, to the bottom.

Watch the audience gape!

Now they believe you have this fantastic control over water, tell them that water will always obey you, and from now on will keep you clean even if you never go near the bathroom again!

Well, you can try.

SO WHAT IS THE 'SKIN'?

Get a bowl of water and try to pick the skin off it. You can't, however hard you try; it's not like the skin on cold custard.

Water is made from trillions of tiny particles called **molecules.** The molecules like to stick together, rather like a group of friends sticks together – they like each other's company, but they aren't completely glued to each other.

The molecules in the middle of the water have friends all around, but those at the surface have no friends above, so they cling extra hard to the ones at their sides. This layer is the 'skin'. It's called the **surface tension** or **meniscus.**

HOW DO WATER MOLECULES STICK TOGETHER?

You can make bits of paper stick together just like water molecules. If you rub a plastic comb or pen on your clothes for a few seconds, then hold it over some small pieces of paper, the paper will jump up and stick to the comb or pen.

Scientists call this an **electrostatic force**. It's actually one of the most powerful forces in nature. I know that picking up bits of paper with a comb doesn't seem very powerful, but remember that in another form, electrostatic force produces thunder and lightning – that *is* powerful!

Those surface molecules in water have a lot of pull. Turn on the tap just a little and see how the surface tension pulls the stream of water into lots of balls.

The No Spill Spell

Let's see just how strong the skin on water really is.

THE EFFECT

Water rises above the top of a glass with no visible means of support, held up by the psychic powers of the magician.

YOU NEED

- Glass of water
- Pile of coins
- Tray to perform the trick in – there may be some spilling

TO PERFORM

Ask an assistant to fill the glass right up to the brim. Make sure that the water is absolutely level with the top.

Now you must both concentrate on the water, and persuade it to rise still higher. Is it persuaded? Try dropping a coin in very carefully. Then a few more. Keep going, dropping in the coins ever so carefully until the water rises over the top of the glass. Have a look at how many coins the glass took. That's powerful persuasion!

WHAT HAPPENED?

The surface tension held the water in. If you look very closely you can see the meniscus clinging to the edge of the glass and bowing upwards and over the top.

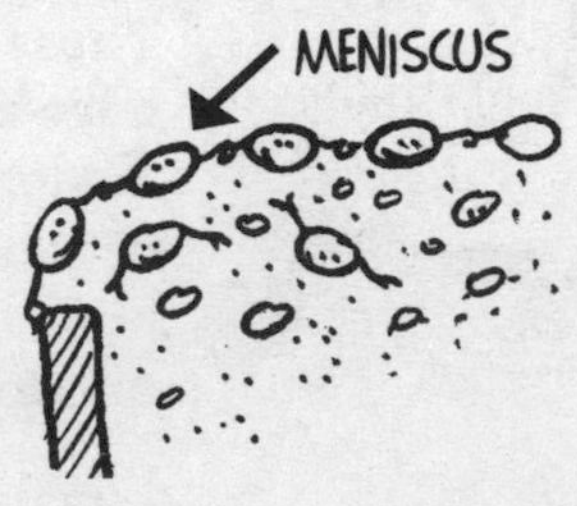

WHY IS WATER WET?

Why is water wet? Seems a daft question. But the answer is even dafter: water is wet because it is sticky.

If you put your arm in and out of a bowl of mercury (a liquid metal), it doesn't feel wet at all; none of the mercury sticks to you. Do the same with a bowl of water and your arm is covered with water. It's the surface tension which makes the water stick.

Those water molecules not only cling to each other, they cling to anything they touch. You can see them clinging on to the side of a glass.

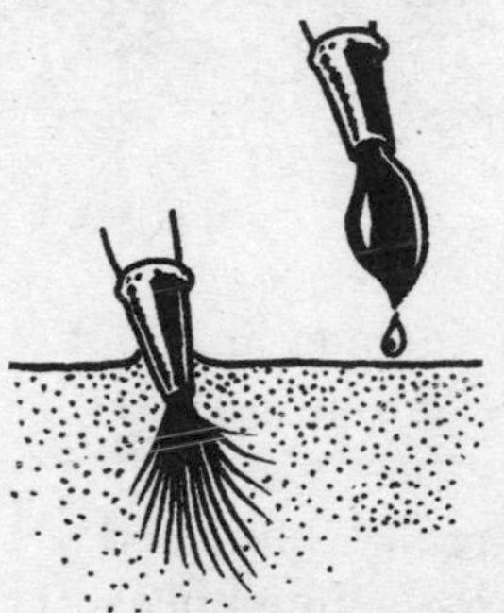

When you put a paint brush into water, all the hairs spread out. But when you take it out, surface tension pulls them into a glob.

If you just touch the surface of the water, it seems to leap up and surround your fingertip.

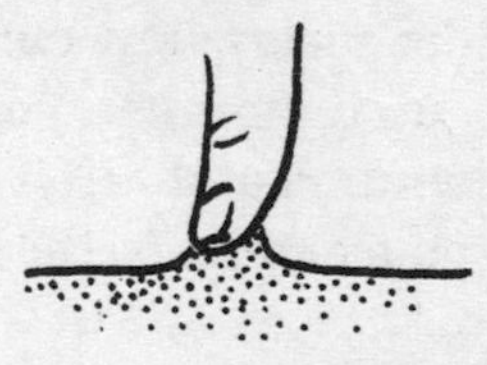

That's why we use water for washing. It clings to all the particles of dirt on your skin and that makes them easier to dislodge.

Here's some magic that uses water like glue . . .

Right Through The Hands

THE EFFECT
The magician can rub woofle dust right through someone's hand.

SECRET PREPARATION
Have a small pile of woofle dust ready in front of you.

TO PERFORM
Choose a victim from the audience. Ask them to hold out their hands so that you can check how clean they are.

> **Misdirection Number One:** You aren't really interested in seeing how clean those hands are. While they hold up their mitts, you casually lick the tip of your right thumb and dip it in the woofle dust. Some of the woofle dust sticks to the tip of your thumb.

Inspect the backs, then the palms. Your victim ends the display with their palms up. You turn them downwards again and tell them to clench their fist tightly, since you are now going to try to make their hands dirty right through to the middle.

Misdirection Number Two: You've already done it! If you turn their hands over as in the drawing, your right thumb presses into the middle of their left hand, leaving some of the woofle dust sticking there.

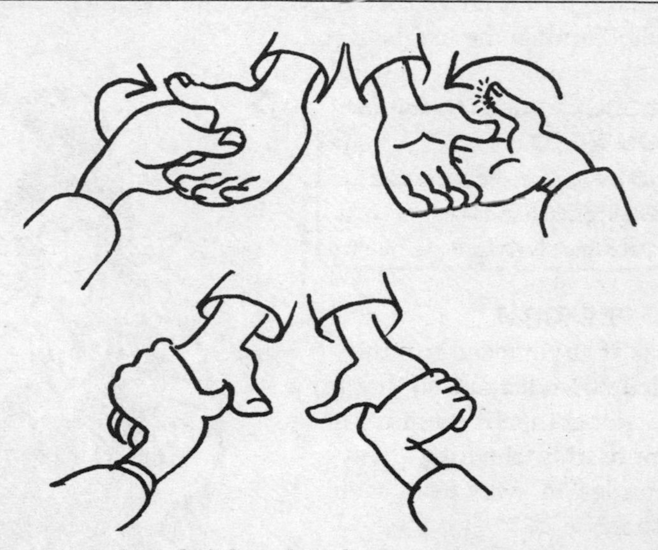

Select your victim's left hand. Take a small pinch of woofle dust and place it on the back of their hand. Say, 'There, I place the woofle dust on the back of your hand, and now if I rub very hard I'll rub it right through to the palm. Keep your hand tight shut so that I can't cheat.'

Rub the woofle dust away from the back of their hand, then ask them to open their hand. There is the woofle dust!

WHAT HAPPENED

The clinginess of water stuck the woofle dust to your finger and then to their hand.

STICKY DICE

THE EFFECT

Can anyone roll two dice down a slope together? Only the magician has the magic touch.

YOU NEED

- Two clean dice, the same size
- Sloping surface, made from a book, mat, board, etc.

TO PERFORM

Ask if anyone can roll the dice down a slope so that they stick together. It seems impossible; the dice always wander off away from each other.

When everyone has had a go, talk to the dice to try to get them to be friends. Go on: lift them up, bring them really close to you and whisper in their ears. While you are this close to them, secretly lick one of their faces. Just a touch is all you need.

Now put them together. The touch of lick will be enough to stick them together while they roll down the slope.

PAPER PICK-UP

THE EFFECT

It isn't difficult to pick up a piece of paper – unless you have a full glass of water in each hand! The magician does so with ease.

YOU NEED

- Two glasses of water, full to the brim,
- Two pieces of paper, about postcard size
- At least two dry surfaces to perform on

TO PERFORM

Ask for a volunteer – anyone who thinks they can pick up a piece of paper without too much struggle. When your volunteer steps up, put one piece of paper on a dry surface and ask them to go ahead and pick it up. Just before they can set off, say, *'Oh, there's just one thing . . . to make it easier for you, I'll lend you my pair of glasses.'*

They'll think you mean spectacles, but give them the glasses full of water, one in each hand. Say, *'Don't spill any water now, will you?'*

With some struggling – and a little spilling of water – they may manage the task. Thank them for their efforts, then show them the easy way with the other piece of paper on the other (dry) surface. Dip the bottom of one of the glasses in the spilt water and touch the paper with it. The paper will stick to the bottom the glass.

WARNING: if the second piece of paper gets wet too soon, it will stick to the table and won't lift off.

What A Corker!

The next magic plays with the surprisingly lumpy surface of a piece of 'flat' water.

THE EFFECT

If a cork is placed in a glass of water, it floats to the edge and stays there. Only the magician knows the way to get it to float in the middle of the glass – bribe it!

TO PERFORM

Fill up the glass practically to the top. Ask a volunteer to try and float the cork in the middle. However carefully they place it, the cork seems magically attracted to the edge.

Tell your audience the cork is sulking because it isn't being paid enough. You will have to bribe it to float in the middle.

Carefully slip coins into the glass, as in the No Spill Spell (see page 14). When enough money has been offered (i.e., when the water level has risen above the top of the glass), the cork will saunter into the middle.

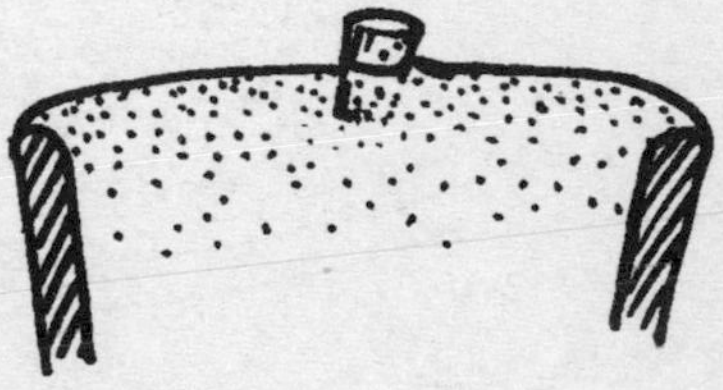

WHAT HAPPENED

The odd thing about this is that in the first case, the cork has to move uphill to reach the edge. When the water rises above the top of the glass the cork seems to move uphill to reach the middle.

Woaah! Aren't we looking at this in the wrong way? The cork floats to the top of the water, right? Well, the top – the highest point – is at the side when the water is down in the glass . . .

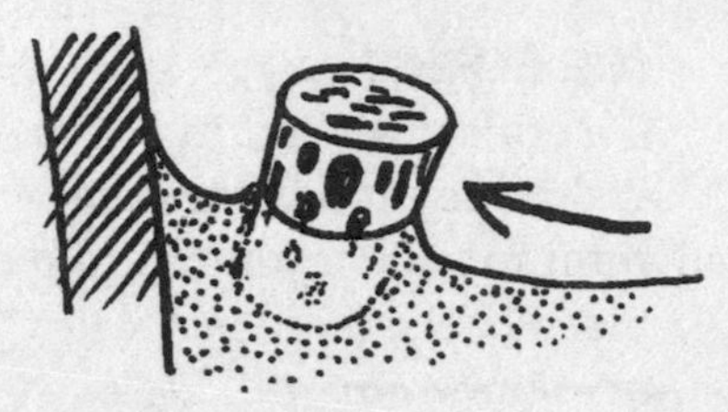

. . . and in the middle when the glass is over-full.

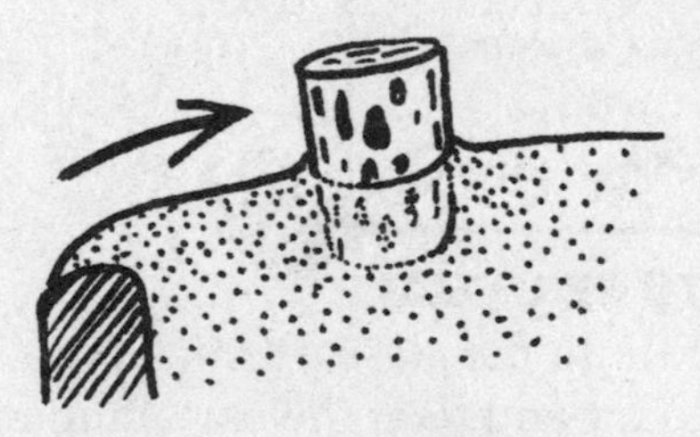

This explains one of the strangest things about breakfast cereals: have you noticed how rice krispies always cluster together in 'rafts'? The surface of the milk is not in fact smooth, but is like a landscape with hills and valleys of meniscus, and the krispies floating to the top of each hill.

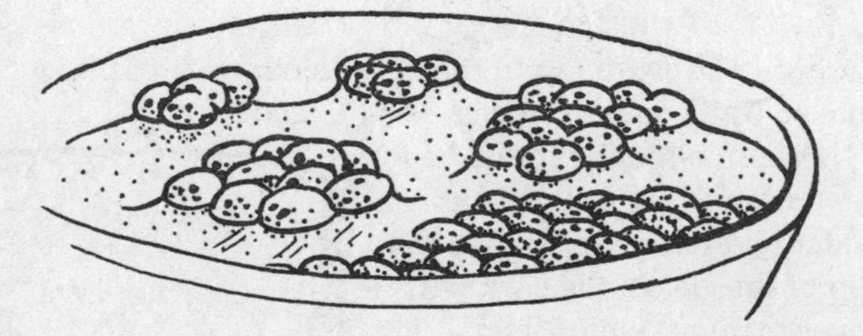

GOING AGAINST THE FLOW

You can make water flow uphill. Put a dry sponge into a puddle of water and see how the water climbs up it. It's hauling its way up the tiny spaces in the sponge.

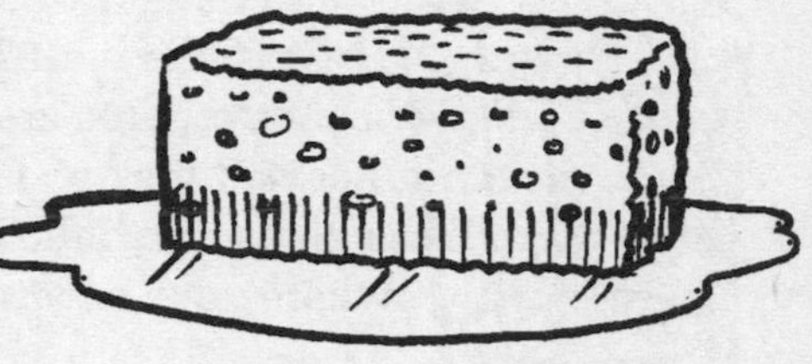

It's the same with paper. Dangle the corner of a piece of tissue paper in some water and see how far it rises.

Paper is made from lots of tiny wood fragments with air spaces between. The water climbs up these spaces.

This is how the towel sucks up your water after the bath; the water climbs up the spaces between the cotton fibres. This ability to climb upwards is known as **capillary action.**

CHROMATOGRAPHY

If you are going to watch water creep up a piece of paper, you might as well make it colourful. Tear a strip off the edge of a newspaper and paint some dots on it with felt-tip pens of various colours. Paint them low down, but not quite at the bottom.

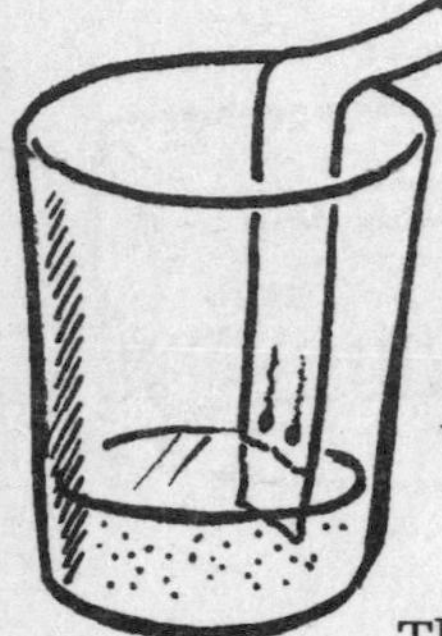

Pour about 4 cm of water into a glass and hang the strip over the edge so that its bottom is about 1 cm under water and the coloured dots are just above the surface. Leave it for one hour, and you will find that a blend of rainbow colours has appeared where your felt-tip pen marks were.

The water crept up the paper by capillary action. When it met the inks it carried them along as well.

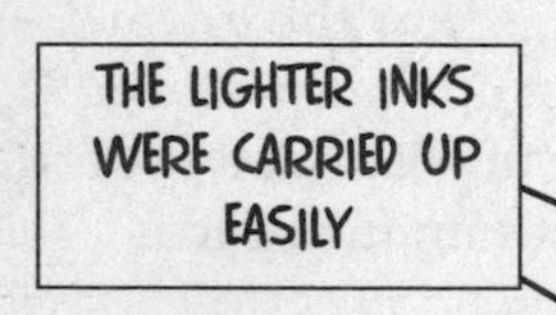

The Now-I-Can-Go-Away-On-Holiday-Plant-Waterer

Capillary action is a helpful in a thousand processes including: the flow of blood around your tiniest blood vessels (known, unsurprisingly, as **capillaries**); the rising of water in a plant from root to shoot; and the lifting of melted wax up a candle wick.

Here is a way to use capillary action to make a cheap device that will solve one of your parents' most pressing problems of all – how to keep potted plants watered when you all go on holiday.

All right, I admit it's not a problem of Earth-shattering importance. However, your parents will be grateful to you for this little nugget.

Fill a glass with water and place it next to the plant you want watering, slightly higher than the top of the pot.

Soak some string and hang it over the edge of the glass so that the drips fall on to the plant. It will keep dripping for days.

Capillary action hauls the water up the string as far as the edge of the glass, then the water drips down into the pot.

CHAPTER TWO

Dishing The Dirt

Even if you sit perfectly still and do nothing all day, you still get dirty. After all, so does the furniture. You just can't avoid it.

The air is full of tiny dust particles, each one of which has a story to tell. Some come from Mars; others from trillions of miles beyond the solar system. There are grains of sand from a dust storm on the other side of the Earth; pieces of lava from the volcano on Montserrat; particles of smoke from a steel works in Romania.

Some particles are pollen grains that dropped off a passing bee. Others are the poos of tiny dust mites that live in the carpet. The dust mites are vanishingly small, and their poos are so minute that the tiniest breeze will lift them up in the air. Some people suffer snuffles and sneezes in an 'allergic reaction' caused by breathing in these poos – I don't blame them.

The dust particles also contain quite a lot of dead vegetation from the garden; a good deal of pieces of household furnishings ; coach-loads of germs and viruses; and large amounts of you.

Yes, YOU. Over half of domestic dust is human skin flakes. We shed our skins all the time in a continuous, slow avalanche of wasted surface cells. To give you an idea how fast this happens, make a mark in ink somewhere you can avoid washing – perhaps just above your elbow. Check it out in two days' time . . . gone! Floating round the house somewhere!

All this dust swirls around you all the time. You can see it sometimes, floating in beams of light shining into a dark room. And of course it settles on you. All over!

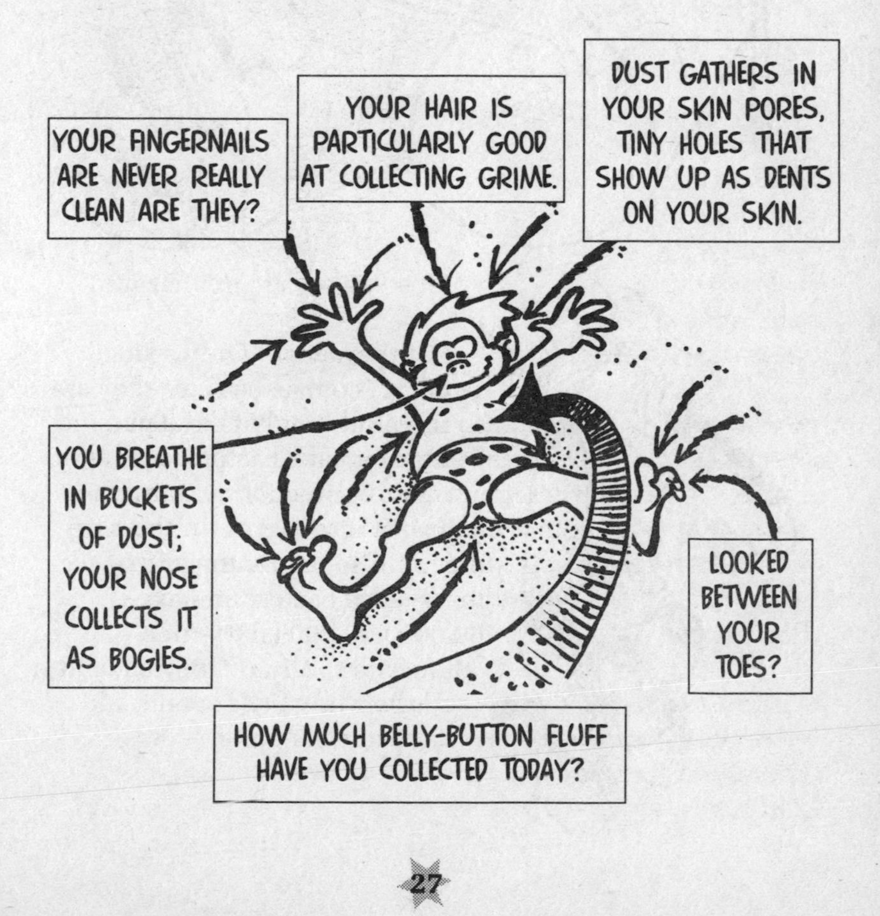

BACTERIA

Some of the bits on your skin are alive! They are germs, or bacteria.

There are a few things you should know about bacteria.

They are small. Very small. So small you can't see them. Here's a collection of them on the point of a sharp needle.

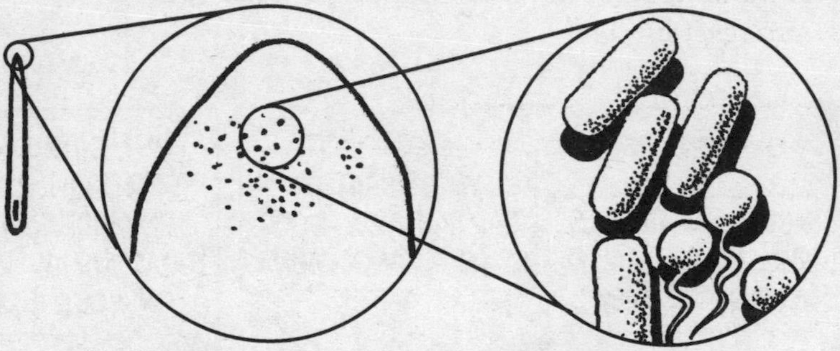

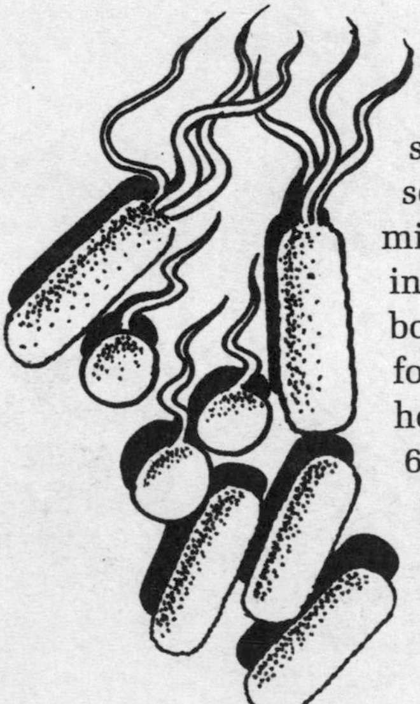

Mostly, they are not harmful.

But they smell! Only a small smell, of course, because they are so tiny. But it only takes ten minutes for each bacterium to split into two. In another ten minutes, both split into two again, making four. And so on. By the end of six hours, each bacterium has 69,000,000,000 (sixty-nine thousand million) children, and all their tiny little smells add up to one big pong!

Paper Increases In Creases

Here's a trick which shows you how quickly one bacterium can become an empire.

THE EFFECT

A humble piece of paper is stronger than the strongest human.

YOU NEED

- One small piece of paper – half an A4 sheet perhaps (you don't have to be precise)
- One large sheet – as big as you can get (a full newspaper sheet will do)

TO PERFORM

Take the small piece and fold it in half as many times as you can in front of your audience.
(You won't be able to fold it more than eight times.)

Now take the big piece and challenge the biggest, strongest, toughest person around to fold it more than eight times. Ask them how many times they think they can fold it – 10 . . . 20 . . . 30?

Watch them struggle as they try to fold it more than eight times. They won't be able to! And it doesn't matter how big a sheet they use.

Every time the bundle is folded in half the thickness doubles, so the thickness increases very rapidly, If you could fold the paper 14 times it would be over a metre thick. 23 folds and the bundle is nearly a kilometre thick. 42 folds, and it stretches 440,000 km into space – farther away than the Moon.

Just think – when your bug population doubles 42 times, you get about four and a half million of them from one mother. Four and a half million mini-pongs equals one stink.

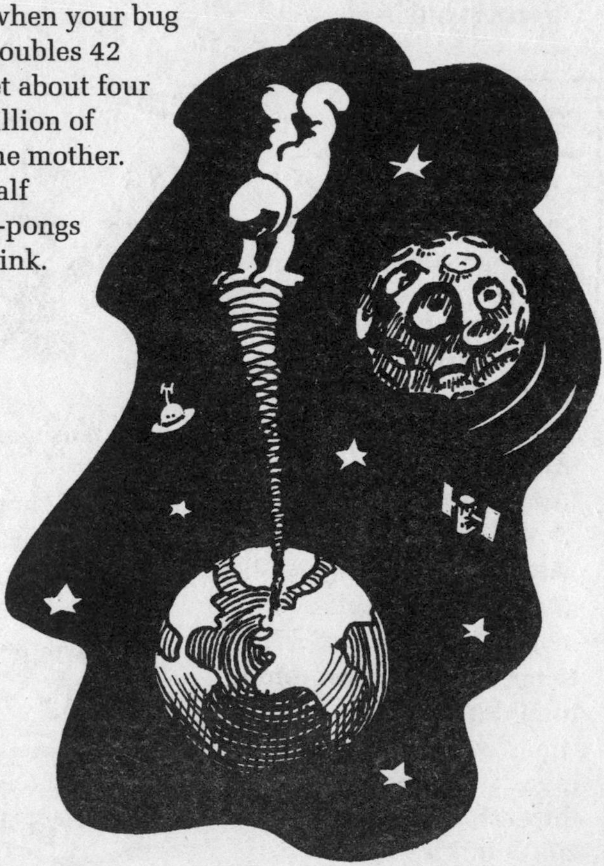

NAPKIN NOBBLING

We've already seen how water grabs hold of just about anything it comes across. Here's what happens when it comes across paper.

THE EFFECT

Paper napkins are surprisingly difficult to tear apart when they're twisted up. But one touch from the magician's magic fingers makes the job deliciously easy.

YOU NEED
- Paper napkin or two sheets of kitchen towel
- Glass of water

TO PERFORM

Roll the napkin up into a tight tube, as tight as can be. Offer it to the biggest person to try to pull it apart. (If by any chance they do manage to snap it, double the thickness and start again.) As they struggle with it, offer to help. Touch the napkin gently in the middle and it will snap.

The trick is to dip your fingers into the glass of water before touching the napkin – the effect of a little water is astounding.

WHAT HAPPENED

So how did a tiny bit of water have such a drastic effect?

Paper is made from mashed-up trees. The wood pulp is laid out in a thin layer and allowed to dry. That's all, there's no glue or anything; so why does it stay together? The answer is that the particles of wood are held together by electrostatic charges like the one which attracted the paper to the comb on page 13. Water muscles in on this, interfering with the electrostatic bonds, and so loosening the wood particles' grip on their neighbours.

This explains the weird behaviour that parents go through when they iron clothes. Having carefully dried their shirts, they get them ready to iron by liberally wetting them again! The water in a steam iron loosens the bonds that formed as the shirt dried, so the creases are easier to remove.

And that's what happens when water meets bits of dirt on your skin. It loosens them and carries them away into the drain. Except for GREASE.

CHAPTER THREE

GREASE

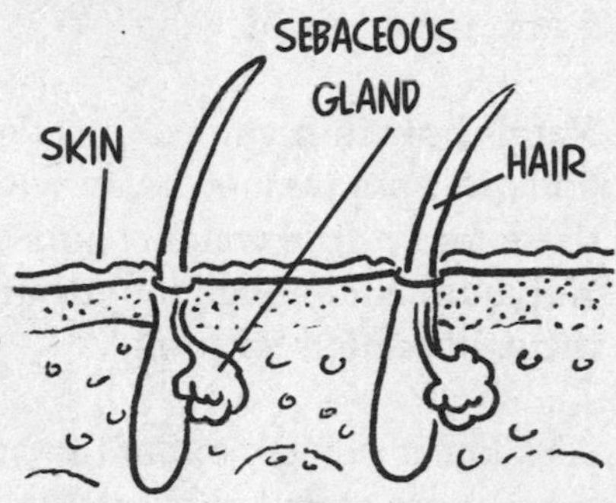

You can't avoid dirt or bacteria, and you'll never avoid grease. It's no good saying, 'All right, I'll keep away from greasy things, then it won't trouble me . . .' because YOU are a greasy thing.

Your body is constantly squeezing oil out of thousands of **sebaceous glands** all over you. Is this some kind of plot to make you greasy out of spite?

Think again. You need grease like frogs need ponds. Remove a frog from its pond, leave it in the sun and it'll dry up very quickly – in fact it'll practically frazzle up into thin air. (Take my word for it; don't try it.) Frogs have very thin skins, so they have to try to keep their body fluids from seeping away all the time. When they're not in the pond, they'll be lurking in the dampest part of the garden they can find.

Human skin is not much thicker, so it comes supplied with oil from the sebaceous glands to help your skin behave more like a mackintosh – but one which keeps the damp in and the dry out.

Here's how thin your skin is. When you have a bath your body oils are washed away with the dirt. If you stay in there for a while water begins to be absorbed by the body. Not a lot, but enough to puff up the tips of your fingers a little. It's called 'pruning'.

As soon as you get out of the bath, your sebaceous glands set to work covering your skin with some fresh oils. Sometimes people use creams or oils as a booster, but normally your natural oils will do all right without help.

We can use our oily skins in a little miracle.

The Miracle Mirror

THE EFFECT

The magician gives a volunteer a maths task to do at bathtime: to think of a three-figure number, do some simple sums with it, and remember the result.

During the bath, their number reveals itself on the mirror, without anyone going near it.

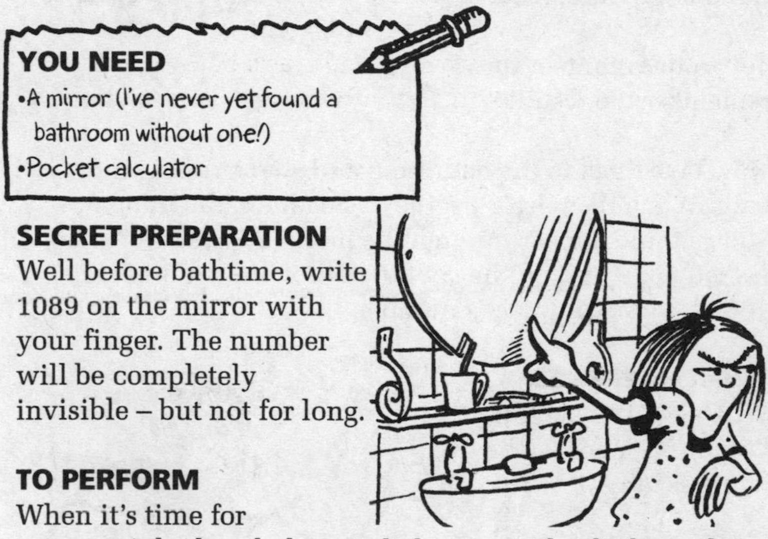

YOU NEED

- A mirror (I've never yet found a bathroom without one!)
- Pocket calculator

SECRET PREPARATION

Well before bathtime, write 1089 on the mirror with your finger. The number will be completely invisible – but not for long.

TO PERFORM

When it's time for someone's bath, ask them to help you with a little maths.

Ask them to think of any three numbers between 1 and 9 (for instance, 4, 2 and 8), then enter them into the calculator as a single number: a three-figure number with the highest figure first.

842

Now reverse the number,

and subtract it from the number on the screen.

Now press equals,

Reverse the number that appears and add that to the number on the screen.

Tell your volunteer to remember the result.

842
\- 248
= 594
\+ 495
= 1089

Now take them to the bathroom and start running the bath. When the mirror begins to steam up, the number 1089 should appear through the mist. (If it doesn't come fast enough, huff on the spot to help it out.) A miracle! 1089 is your volunteer's number.

WHAT HAPPENED

Whatever number your volunteer chooses, the final figure is always 1089.

But what about that finish? The number was written in the natural sebaceous oils at the tip of your finger. The oils didn't allow the mist to settle on the number in the same way as on the glass, so it showed through.

SOAP

Water is pretty good at removing most dirt, but there's one thing it won't touch; it utterly refuses to mix with grease. If you don't believe me, put some cooking oil and water in a jar together and shake them up. They quickly separate out into two layers.

However, there is a way round this; put some washing-up liquid in the jar as well. Now, when you shake them up, the mixture stays mixed. Scientists call this an **emulsion.**

The clever thing about washing-up liquid (and soap) is that it likes to cling to both water and grease. Once it is attached to the grease at one end and water at the other, the water can wash it away and the grease comes along for the ride.

As soon as the water molecules start clinging to the washing-up liquid or soap molecules, they stop clinging to each other. The meniscus skin is weakened.

Let's use this in some more tricks . . .

Soap Boat

THE EFFECT

The magician has made the smallest outboard motor in the world. Attaching it to a cardboard boat propels the boat across the bath.

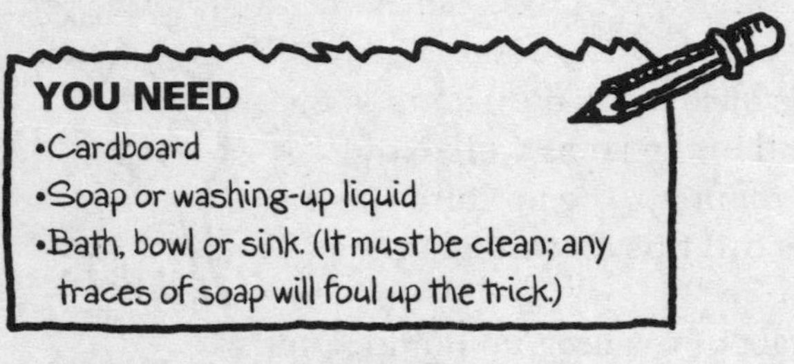

SECRET PREPARATION

Make a simple boat shape out of cardboard, like this:

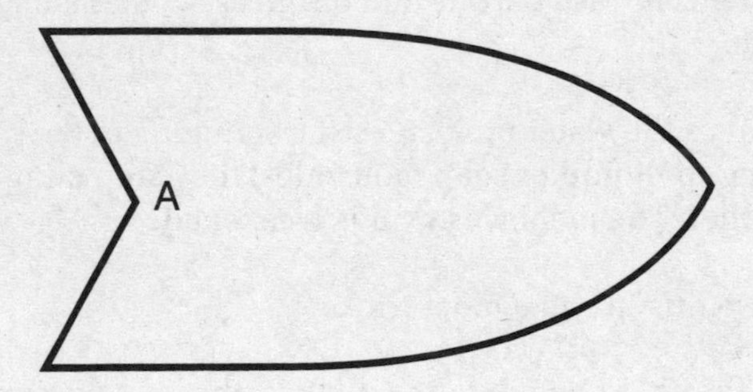

Then fill the bath with water, and secretly put a dab of wet soap or washing-up liquid on the tip of one finger.

TO PERFORM

Introduce your audience to the boat. Invite them to float the boat on the water to check if it is seaworthy. The fact that the boat just sits there also proves that it needs an engine.

Tell them you have designed the smallest outboard engine in the world – so small that it's nearly invisible. Pretend to take it out of your pocket. Point out the elegant shape, the powerful design, the pretty colours. (This is obviously the Emperor's new outboard motor.)

Now take the imaginary engine from them and attach it to the boat. Up to now you have kept your soapy finger away from everything. But as you mime the attachment, touch it against the top of the V cut (point A in the picture). Now, when you put the boat in the bath it will shoot across to the other side, to the amazement of all.

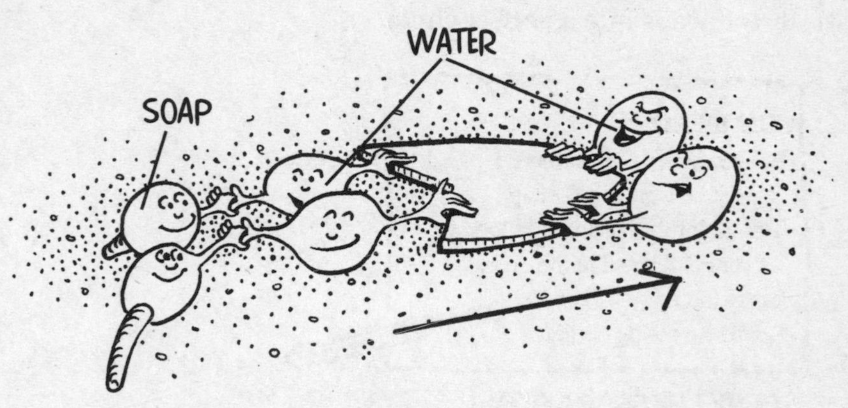

WHAT HAPPENED

Before the soap came into the picture, the boat was pulled equally in all directions by the surface tension of the water. The soap weakened the water's hold on the back of the boat, so the meniscus at the front was able to win the tug-o-war.

The Perfect Circle

This uses the same science as the previous trick. Can you work it out?

THE EFFECT

The audience is shown a bowl of water and asked to try and float a piece of thread on it in a circle.

They struggle with the impossible task until the magician shows how it's done. With a touch of the magic finger the thread leaps into a perfect circle.

YOU NEED

- Bowl or sink full of water – it must be clean of all soap
- A length of thread about as long as your hand, tied at the ends into a loop
- Soap or washing-up liquid

SECRET PREPARATION

Put a dab of soap on the tip of one finger.

TO PERFORM

Invite the audience to float the loop on the water in a perfect circle. They may manage some pretty impressive loops, but mock them all. With any luck your sarcasm will needle them into saying, 'Well you try and do better!'

Simply touch the water inside the loop with your soapy finger. The thread will leap into a beautiful, flawless, perfect circle.

WHAT HAPPENED

As with the boat in the last trick, the surface tension was at first pulling on all sides of the thread equally. When the soap appeared inside the loop, the surface tension was weakened. Meanwhile, outside the loop, the water kept on hauling away at the thread, pulling it outwards equally in all directions to form a perfect circle.

The same equal tugging makes soap bubbles and water droplets round instead of square.

The Fiendish Fetid Finger

THE EFFECT
The magician's finger answers the question, 'Is it time to have a bath?' by touching a bowl of water with pepper sprinkled over it.

YOU NEED
- Bowlful of water
- Pepper
- Dab of soap or washing-up liquid on the tip of one finger

TO PERFORM
Tell the audience you can find out if it's time to have a bath, using the mystic kitchen oracle. Say that if you need a good wash, the pepper will tell you by running from your dirty finger.

Sprinkle pepper on the surface of the water. Touch the middle with your soapy finger, and watch the pepper flee in disgust.

WHAT HAPPENED
The tug-o-war was won by the same science as in the last two tricks.

De-Mistified

This isn't a magic trick, but it is a sure-fire way to gain a bit of credit with your parents!

To stop car windscreens misting up, put a tiny amount of washing-up liquid in a bowl of water, dampen a cloth with it, and wipe the windscreen on the inside.

Can you think why it works?

The answer is below.

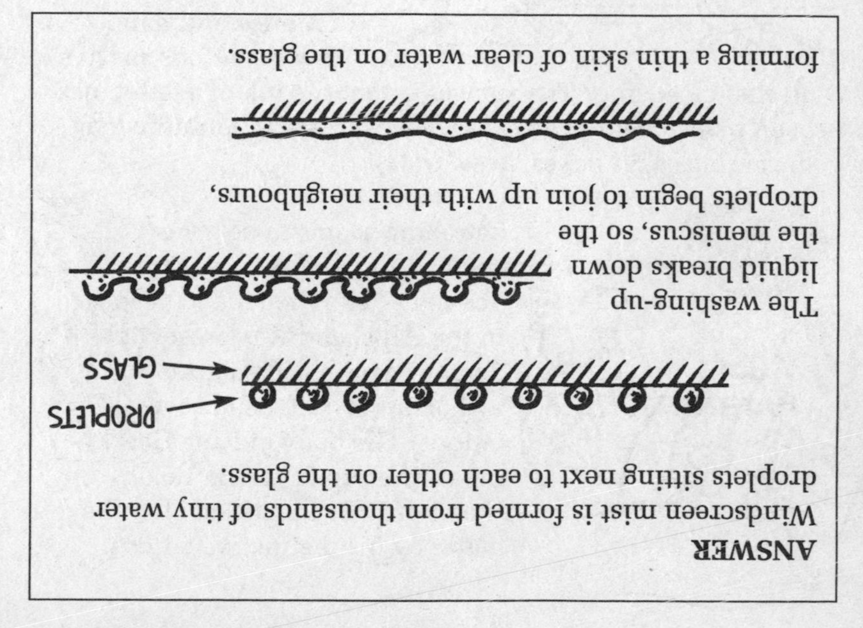

CHAPTER FOUR

COMING CLEAN

STAINS THAT STAY

Some ink stains on your skin seem to stay put however much you scrub them. Don't worry! As you know, your skin is constantly flaking off. It goes at an incredible rate. You replace all your skin every nine weeks, so that stain should be gone in no time.

But there are some everlasting stains that don't wash out – stains that are put there on purpose, at great cost of money and pain. How do tattoos survive all that cleansing? The answer is that the ink of a tattoo has been placed below the skin level, so it stays behind when the surface skin flakes away.

Tattooing seems to be a very ancient and common practice. The 'ice man' who died 5,300 years ago in the Alps, and was preserved deep frozen in a glacier until his discovery in 1991, had several tattoos. The body of King Harold, the runner-up in the Battle of Hastings, was identified after the battle by his distinctive tattoos.

TOOTH TRUTH

Do you like sugar? You aren't alone. The bugs that live on your teeth adore it. But all they give you in return is an acid which eats into your teeth. This is **tooth decay.**

INSTANT TOOTH DECAY

This will show you why you need to clean your teeth. Teeth are made from the same stuff as eggshells. Leave an egg in a cup of vinegar for three days and the acid in the vinegar will dissolve the shell completely. That should get you running for the toothpaste.

WHAT IS TOOTHPASTE?

Would you like to know what's in your toothpaste?

Chalk – to scrape at the surface of your teeth

Paint – to make them look nice and white

Seaweed juice – to give it all a nice gloopy feel

Washing-up liquid – to make the foam

Disinfectant – to kill bugs

And of course, lots of flavouring, so you can't taste the chalk, paint, seaweed, washing-up liquid and disinfectant.

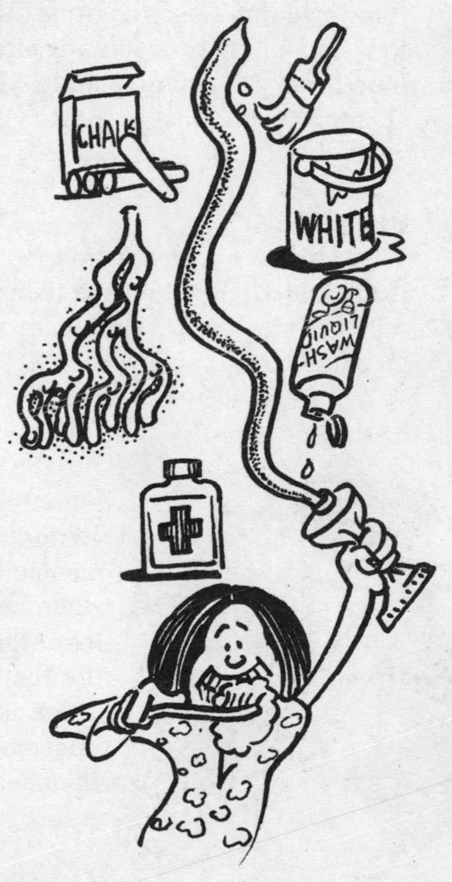

HAIR CARE

The sebaceous glands that help to keep your skin oily also coat your hair. This coating is a natural protection. If you stop using shampoo for a while your hair will keep itself pretty clean . . . well, clean, but not pretty; it will look greasy, of course.

Shampoo, (which is basically washing-up liquid with added froth and perfume) cleans off the grease, and conditioner replaces some of it. Many people around the world don't use shampoo. They have to spend a long time looking after their hair. Mind you, many people who do use shampoo spend just as long.

BIG WIGS

Wigs are more widespread than you think. Mostly they're used to cope with the strange fact that many men go bald.

Why should so many men go bald? Perhaps the reason will be found over the next twenty years, but for the time being there isn't a scientist anywhere who knows the answer. They're all worrying themselves bald thinking about it.

CHAPTER FIVE

Light Delight

The next few tricks should be done carefully. In fact everything you do in the bathroom should be done carefully – especially bathing.

Bathrooms never seem exactly user-friendly, do they? Look at all those cold hard, ceramic surfaces, waiting for you to bang your head against them. See all those puddles of water waiting for you to slip on them, to make head-banging so much easier. Notice those smooth surfaces, apparently designed to make the puddles as slippery as possible for the head-banger. Where's the sense in it?

In fact bathrooms are the way they are because water is not welcome in the home. When things get wet they rot. Pounds and hours are spent every year keeping water out – mending roofs, damp-proofing foundations, sealing windows, etc. So if water is to be invited in, it has to be carefully supervised between leaving the tap and entering the drain. All those tiles and wipe-clean surfaces are designed to keep the enemy under strict control.

Another hair-raising quality of water is its ability to give you an electric shock. Not by itself, of course, but it can if it gets inside an electric appliance. Water carries electricity very well. So if water gets inside a piece of electrical equipment, and if you're somehow connected to the same piece of water as the appliance, then they'll be peeling you off the ceiling. Even light switches are located either in the ceiling or outside the bathroom door to keep them out of harm's way. The message here is: NEVER take electrical appliances into the bathroom.

Water also makes glass disappear from view; and if it's broken glass, then that spells another kind of danger . . .

All Gone Glass

Put a small glass, such as a tumbler, inside a pint mug. Look through the sides and you can see the tumbler clear enough. Now fill with water and look again. The tumbler has virtually disappeared.

WHAT HAPPENED

We see objects because they reflect light waves, some of which find their way into our eyes.

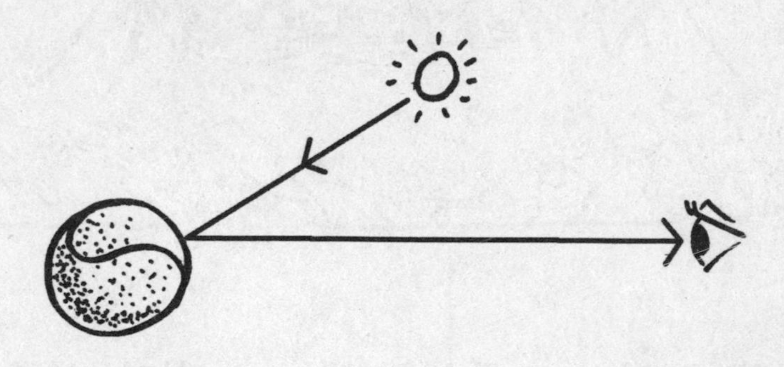

Glass is transparent, and so is water – light waves shoot straight through. So how come we can see drinking glasses and water in swimming pools? Surely they should be invisible?

Luckily for us, light is slowed down by water and glass. The result of this is dramatic.

Imagine a line of children running across a field holding hands.

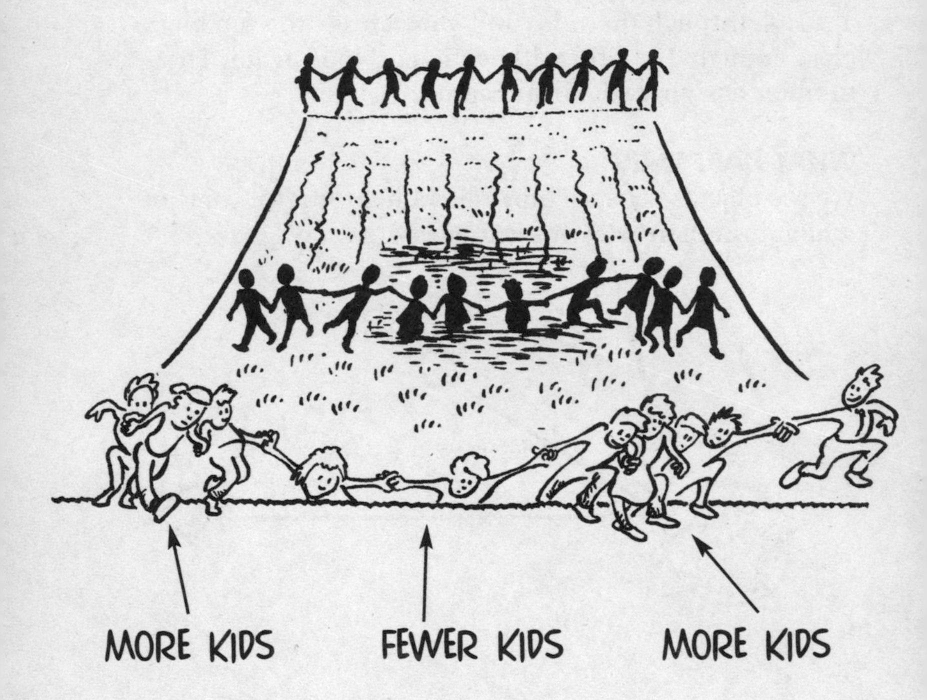

Some of them have to splash through puddles, which slow them down a little.

Some have to wade through lakes of mud, which slow them down a lot.

By the time they get to the other end, the organized line has become a jumble.

Light waves have the same problem with water and glass. Water slows them down. Glass slows them down more because it is denser than water.

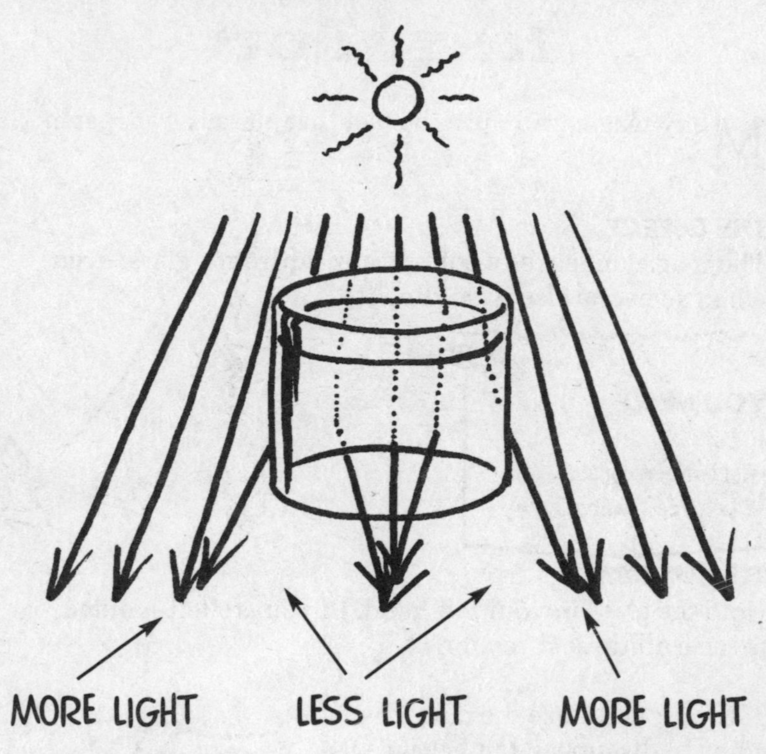

So waves of light travelling through a drinking glass are twisted, or **refracted**, by the time they reach our eyes. The same is true for light travelling through water.

But put glass into water and, because they both refract light by about the same amount, the glass disappears. A glass of water looks about the same as a glass of water-and-glass.

And that's why you can't see broken glass in the bath. You have been warned!

That's The Way The Money Goes

Many magic tricks use the fact that glass is transparent.

THE EFFECT

The magician makes a coin disappear from a glass, even when someone else is holding it.

YOU NEED

- Coin
- Flat bottomed glass
- Cloth, tea-towel or hanky

TO PERFORM

Hold the glass in your left hand. (If you are left-handed, reverse all these directions.)

Tell your audience the coin is going to disappear, but because it's a shy coin, it'll do it in private.

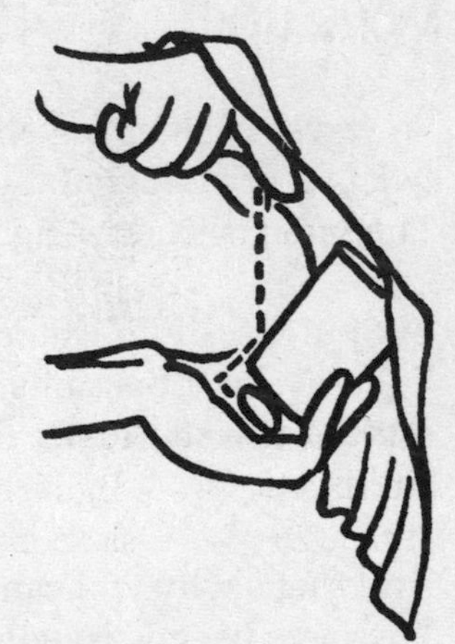

Cover the glass with the cloth. Holding the coin in your right hand, make as if to drop it into the glass, under the cloth. But don't actually drop it into the glass. Drop it so that it chinks against the outside of the glass, then falls into the palm of your left hand, as in the picture.

Ask someone to check that the coin is still there. Lift the cloth briefly; they'll see the coin through the glass, and because they heard the chink they'll believe it's inside the glass.

Replace the cloth, take the glass in your right hand, and ask them to hold it. As you pass the glass to them let your left hand drop casually to your side, with the coin held loosely in it.

While they hold the glass, remove the cloth and reveal the glass – empty!

WHAT HAPPENED

In this trick the light from the coin travelled straight through the glass. In the next tricks, things get wilder.

That's The Way The Money Comes Back

A snazzy follow-up to 'That's the Way the Money Goes', this one uses light refraction to fool the audience.

THE EFFECT

The magician makes a coin suddenly appear inside an empty glass.

YOU NEED

- A glass full of water – it should have a wide, flat base
- Saucer or bowl
- 5p, 10p or 20p coin

SECRET PREPARATION

Make sure the coin and the base of the glass are both dry.

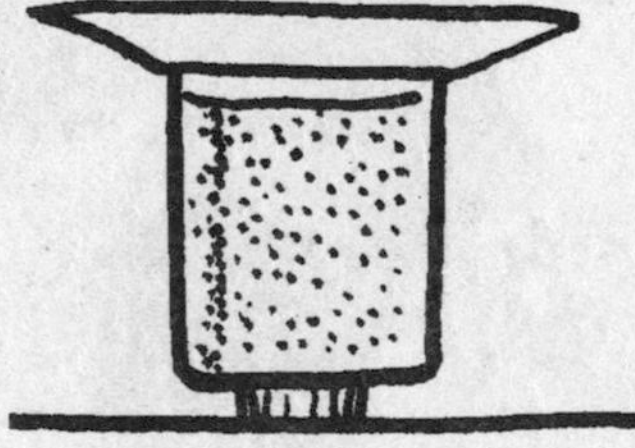

Place the coin on the table. Place the glass of water on top. Place the saucer on top of the glass. The coin is now invisible, however hard you look for it.

TO PERFORM

The glass/coin set-up can be left anywhere. When you think you'd like to make money appear, show the audience the 'empty' glass. They must check very carefully that the glass is really empty (without touching). Wave your magic fingers over the glass. Remove the saucer and invite them to look in. There's the coin!

WHAT HAPPENED

When you look from the side you ought to see the coin, but the light from it has been refracted upwards by the glass

and water. You could have seen the coin from higher up, were it not for the saucer blocking your view. As soon as the saucer is removed, the coin comes into view.

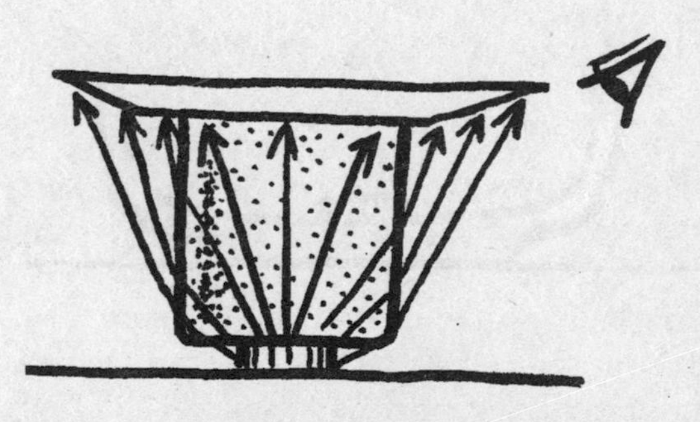

That's The Way The Money Floats

More mind-bending, light-warping money manipulation.

THE EFFECT

A coin is placed at the bottom of a dish. When water is poured into the dish, the coin seems to levitate.

TO PERFORM

Place the coin in the bowl. Ask a volunteer to position themselves so that the coin is just hidden by the rim.

Hypnotise your volunteer. This is done by dangling your fingers in front of their face and saying slowly, *'You are feeling sleepy . . . you are feeling ve-e-e-ery sleepy. You are falling into a deep trance. You imagine that you see the coin floating upwards . . . upwards . . .'*

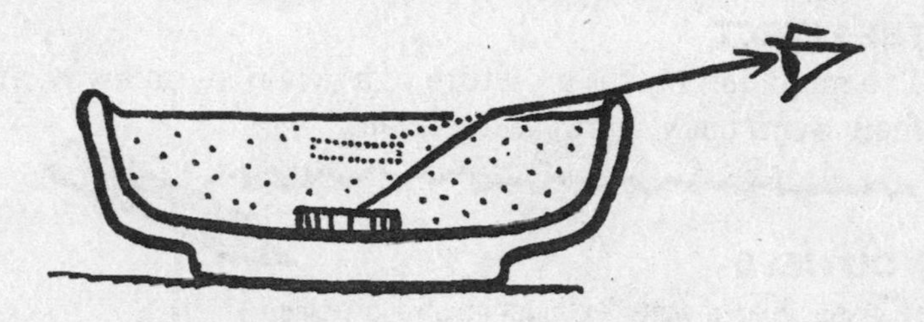

While you say all this, pour the water gently into the bowl. Your volunteer will see the coin rise above the rim of the bowl.

Snap your fingers, and they can check that the coin is still on the bottom of the bowl.

WHAT HAPPENED

The water bent the light from the coin so it seemed to move.

Well Swum, Swan!

This light-warping caper can turn a picture round completely.

THE EFFECT

The magician makes a picture of a swam swim away, and then swim back again, all by itself.

YOU NEED

- A glass or jar of water – it must be straight-sided
- Small piece of paper with a picture of a swan on it, about half the width of the glass (see picture)

TO PERFORM

This is another one-to-one trick. It goes with the old tongue-twister:

> Swan swam over the sea.
> Swim, swan, swim!
> Swan swam back to me;
> Well swum, swan!

Position your subject so that they are looking square on to the glass. As you recite the first two lines, 'swim' the picture towards the glass. As you get to the second part, (about swan swimming back), let the swan pass behind the glass, about 5 cm behind it. Your subject will see the image reverse – the swan appears to be pointing the other way.

WHAT HAPPENED

Because of the curvature of the glass and water, the light coming from the picture crossed over – the left edge seemed to be on the right side, and vice versa. (Magnifying glasses turn the image back to front and upside down.)

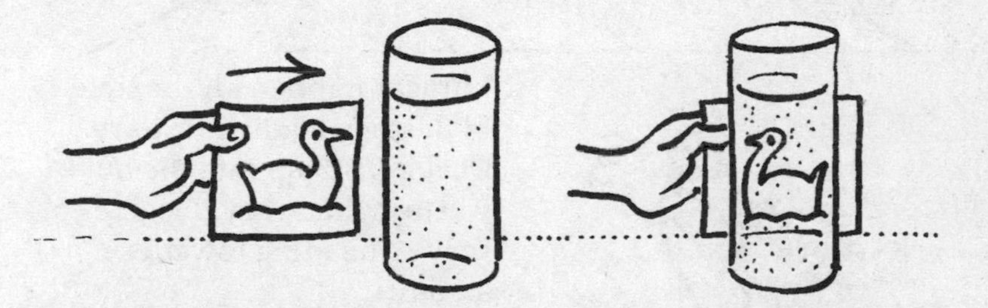

Even air refracts light a little. Warm air is thinner than cool air, so light travels at slightly different speeds. The difference is minute, but the eye can spot it. Look across the top of a toaster when it's in full swing. Objects in the distance seem to shimmer. That shimmer happens because some of the light travelling to your eye is speeded up by one hundred millionth of a second. Not exactly a gigantic difference, but on a hot day you can see a strange side-effect of this . . .

MIRAGES

If you are driving along a straight flat road on a hot day, look out for puddles in the road ahead. They'll appear far away near the horizon, then disappear as you get closer. These are **mirages,** caused by light from the sky being refracted by the hot air just above the tarmac. Your eye may be looking at a patch of road, but it sees a patch of sky instead. Your brain comes to the only conclusion it can under the circumstances: it decides there's a puddle in the road, reflecting the sky.

Mirages happen all the time in deserts, leading unwary travellers deep into the hot wastes when they imagine they're heading towards an oasis.

Magic

Invisible Ripple

Fill a bath with water and look at the way the light plays on the side of your bath below the water's surface. Now flap your hand under water and see the ripple in the light patterns. The ripple is invisible on the surface, but your flapping has compressed parts of the bath water, making it denser, so it refracts the light, hence the light show.

Magic

Topsy Turvy Tootsies

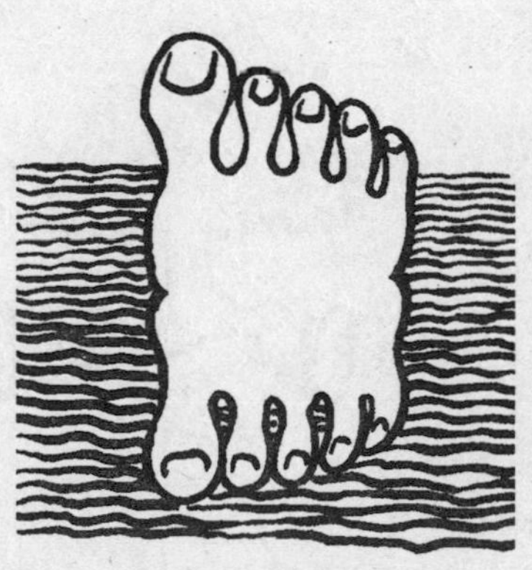

Lie in the bath with your toes sticking out of the water. Instead of being refracted by the water, the toes are reflected by it.

If light hits water at a very flat angle it bounces off it, in the same way that stones bounce off water when they are skimmed across a pond.

You can do magic with this next time you go shopping. . .

Shop Window Acrobatics

Have you noticed that shop windows turn into mirrors if you look along them rather than through them?

Find a plate-glass window with a corner on it, then stand with the corner running straight down the middle of you, as in the drawing. Ask a viewer to stand close to the glass.

Lift your outside leg, and it will seem that you are floating.

CHAPTER SIX

AIR WE GO! AIR WE GO! AIR WE GO!

Air can be just as strange as water, and these tricks make use of air's unexpected properties. They also make a bit of a mess, so the bathroom is the best place to practise them.

Magic

AIR IN THE WATER

THE EFFECT

The magician can hold a handful of tissues under water without them getting wet.

YOU NEED

- Tissues
- A glass
- Sink or bath with water in

TO PERFORM

Keep the glass hidden to begin with.

Claim that you can hold a handful of tissues under water and keep them dry. Gather a fistful of tissues, produce the glass, ram the tissues up to the end and thrust the glass under water, upside-down. The tissues remain dry.

WHAT HAPPENED

You trapped a bubble of air inside the glass. The water couldn't force its way in.

WATER IN THE AIR

THE EFFECT

The magician pours a glass of water over someone's head . . . but the water doesn't come out of the glass.

YOU NEED

- Glass of water
- Postcard

TO PERFORM

The magician asks if anyone would like to be king.

Pick a volunteer and say, '*Step forward. I will anoint you King . . . King Wet . . . SoaKing Wet. Here is the glass of holy bath water to be poured over your head.*' Produce a glass of bath water. '*Here is the Card of Truth.*' Produce the postcard and place it on top of the glass. '*If you are truly born to be king, then the holy waters will pour all over your head. If not, the water will remain in the glass.*'

Hold the card in place while you turn the glass over and move it over the victim's head.

Say, '*I will now release my hand.*' Do it. The water will remain in the glass.

'*You are obviously an imposter. Off with his head!*'

WHAT HAPPENED

This is one of those crazily simple tricks that never fails to amaze even the most hardened scientist. What it demonstrates is the weight of air pushing in on you from all sides.

If I asked you to walk about with a grown-up on your head all day long you'd be pretty uppity at the thought. Yet that is what you already do. That is the weight of air that presses down on you from the 80 km thick atmosphere above your head. You're so used to this task that you don't notice. But that pressure is there all right.

That same **air pressure** pushes against the card on the mouth of the glass.

The water in the glass wants to drop out, of course, but as it starts to drop it creates an empty space above. If it's going to come out, something else will have to go in. And there's the problem. Nothing can get past the card across the mouth of the glass, pushed up by air pressure. So the water has to stay put.

Boiling Cold

With this trick you find a way to allow air into the top of that glass.

THE EFFECT

The magician makes a glass of water boil on command.

SECRET PREPARATION

Lay the hanky over the top of the glass. With your finger, poke it halfway into the glass. Fix it near the top with the elastic band.

When you turn this lot over the excess water will pour away, but the water inside the glass will remain. The hanky, meanwhile, will arch inwards in a beautiful curve.

TO PERFORM

Hold the glass around the elastic band. Show the audience the glass with water in it. Ask them to touch the glass to prove it is cold.

Now make the water boil. Place a finger of the other hand on the base of the glass. Concentrate horribly and push downwards with the finger, while slightly easing your grip on the hanky.

The glass will slip downwards. As it does so, it will pull the hanky out of the glass. Bubbles will appear in the glass and it will look as if the water is boiling. To add to the effect, crease your face in pain and say, '***Oooooh, the heat!***' Then stop pushing, breathing a sigh of relief that the water has gone cold again.

WHAT HAPPENED

The water was held in there for the same reason as in the last trick. Although there are holes in the hanky material, they are so small that the water's surface tension wouldn't allow air through. As you pulled the hanky out of the glass, however, you forced the issue. You created so much empty space, or **vacuum**, inside, that air pressure on the outside forced air up through the holes. Those were the 'boiling' bubbles.

POOR POURING

Once more the magician can control when water comes out of something by controlling when air goes into it.

THE EFFECT

A perfectly ordinary looking bottle of water suddenly sprouts leaks everywhere.

WARNING: This is a messy trick; be very careful where and when you choose to do it.

YOU NEED

- Bath with water in it
- Large empty cola bottle
- Pair of scissors and grown-up
- Water

SECRET PREPARATION

Ask the grown-up to take the scissors and carefully poke four small holes around the bottom of the bottle and one right at the top (see picture).

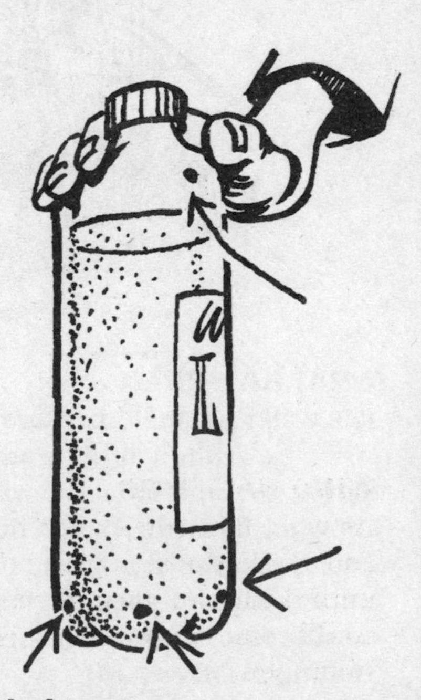

Fill the bottle with water by dipping it in the bath, then screw the top tightly on. As soon you lift the bottle out of the water it will start to leak. Hold it so that you can cover the top hole with your thumb. With the top hole covered, the leaking stops.

TO PERFORM

Simply ask someone to hold your bottle for a minute. Tell them you've just got to pop off and find a towel. As soon as you let go, they'll know why you wanted the towel – water will pour everywhere.

WHAT HAPPENED

As with the last trick, the holes in the bottle were too small and surface tension was too great to let air in or water out, until the hole at the top was released. As soon as the air could come in, gravity got its chance. It took over, and the water poured out.

Magic Drink Pourer

This is a slightly more civilised way to do the last trick. This time the grown-up-with-scissors makes the same kind of hole at the top of the cola bottle, but only one at the bottom. At meal-time fill it with the juice of the day, being sure to keep a finger over the hole at the bottom until the top is screwed on and your thumb is over the hole at the top. Now the bottle won't leak, so you can lift it with confidence, carry it to the first glass, and hold it ready to pour.

Say '**Pour!**' and at the same time move your thumb a tiny bit, letting the air in and the juice out.

When the glass is full say, '**Stop!**' Cover the hole again and move on to the next glass. (If you like, you can get everyone to say '**when**' to the bottle when it has poured enough.)

CHAPTER SEVEN

HOT AND COLD

It took us humans sixty-five million years to evolve from little hairy mammals hopping between the toes of dinosaurs to the magnificent specimens we see in the mirror today. But somewhere along the way we lost something – fur. How did that happen? Surely it's a mistake. So much of our time is spent trying to keep warm, making clothes and fires etc. If only we hadn't lost our fur none of this would be necessary.

Humans are only happy in a very narrow band of temperatures – between 34°C and 38°C. Most of the time you control your body temperature by putting on suitable clothes. But there comes a time – bathtime – when you have to fall back on your natural, rather ropey techniques of temperature control, or **homeostasis**.

TOO HOT?

You sweat to try and remove the heat by evaporation. The heat turns your sweat into water vapour.

You go red because your body automatically opens up the blood vessels near the skin's surface, to try to lose heat.

You also run around less because doing exercise makes you even hotter.

TOO COLD?

You huddle, to try to keep as little of you exposed to the cold as possible.

You go pale, as the blood vessels near the skin are closed off the keep the heat in.

You shiver and your teeth chatter. Your body knows that one way of warming up is to do some exercise: shivering and teeth chattering is a special automatic exercise.

You get goosebumps. Millions of years ago when your great-great-great grandparents were covered in hair, tiny muscles made their hairs stand on end when it was cold (in the same way that birds puff their feathers out in winter). Nowadays the muscles are still there, even though the hair has gone. Goosebumps remind us of our hairy ancestors.

One way or another, these little tricks allow you to adapt to the temperature around you.

And a magician can do a trick to fool you.

Hot, Cold And Middling

In this trick the body is fooled; it doesn't know whether it should adapt to hot or cold conditions.

THE EFFECT

The magician can change the temperature of a bowl of water, just like that.

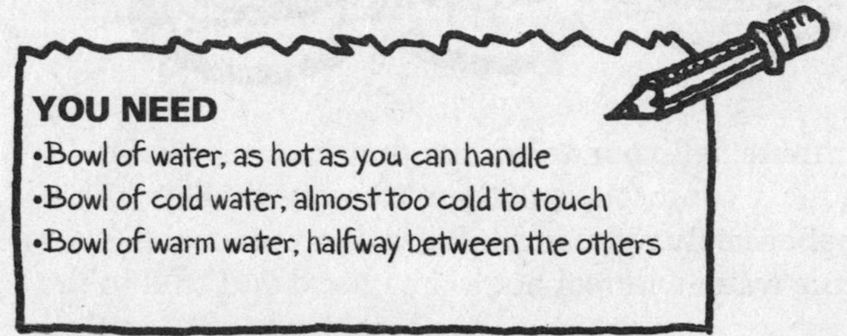

TO PERFORM

While a volunteer rolls up their sleeves, tell your audience about your magic powers to make water any temperature you like.

Bring out the cold bowl, commanding it, '*You are cold . . . You are very cold!*' Ask your volunteer to put one hand in the cold bowl and keep it there, to make sure it doesn't disobey orders and turn warm.

Bring out the hot bowl, and likewise say, '*You are hot . . . you are very hot!*' Ask your volunteer to put their other hand in the hot water and keep it there. Leave them like that for a minute.

Meanwhile bring out the warm bowl. Command it, '*You are hot . . . you are cold . . . you are very hot . . . you are very cold . . .*'

After a minute, tell your volunteer that the warm bowl will be utterly confused. Ask them to prove this by dipping alternately right and left hand in the bowl. Sure enough the water will feel hot to one hand and cold to the other.

WHAT HAPPENED

One hand has adapted to cold water; it now thinks the warm water is hot. The other hand has done the opposite. You'll notice that one hand is pinker than the other – homeostasis at work!

If ever you go to the local swimming pool you will find that moving from the cold pool to the hot pool and back is equally strange. The temperature can seem to swing madly, but it's actually your body that's going mad.

SAVED BY THE TOWEL

When you get out of a hot bath the air seems cold – you reach for the towel! Since nature has done away with your fur coat, towels will have to do.

The towel does three things:

1 It sucks up the bath water that is still stuck to you (by capillary action – see page 23). This is very handy, since that bath water looks suspiciously like sweat; and sweat, as you know is going to help you cool down rather than warm up.

2 It acts as a barrier, stopping draughts from blowing away what heat you have.

3 It **insulates** you. Inside your towel you have your own pocket of air, which your body warms up and keeps snugly wrapped around you. Air is a good insulator. It transmits heat very slowly, so your body heat stays put.

Insulators also keep cold things cold, and the next trick is a tasty way to show it.

Alaska Surprise

Use insulation to make a toasty hot pudding with freezing ice cream in the middle.

WHAT YOU DO

Pre-heat the oven to Gas Mark 8 (450°F / 230°C).

Put a simple sponge base in a baking tray.

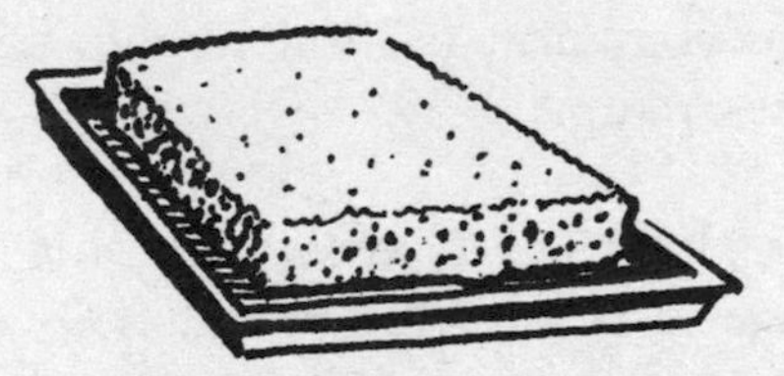

To make the meringue, beat the egg whites until they are fluffy with lots of air, then mix in the sugar carefully, so that the egg white remains fluffy.

Spoon the ice-cream on the sponge base and cover the whole lot with the meringue, making sure it is well covered all round. Then put it in the oven for three or four minutes.

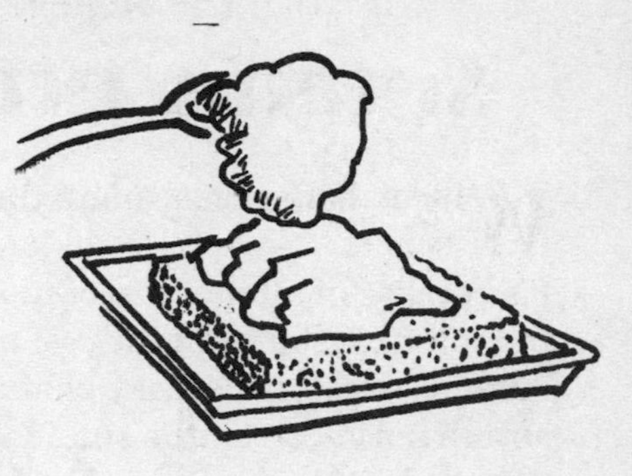

When you take it out, the meringue will be crisp and hot, but the ice-cream will be freezing cold.

WHAT HAPPENED

The air which you mixed into the meringue when you beat it, and the air bubbles in the sponge, all acted as insulation against the oven's heat, so while the outside warmed up, the inside stayed chilly.

CHAPTER EIGHT

SOUNDS INTERESTING

Why do bathrooms sound different?

It's because of all those easy-to-clean surfaces. Without all the soft furnishings in the rest of the house – carpets, curtains, sofas, beds, easy chairs – there is nothing to absorb noises, so all the sounds come back at you. The place echoes with your voice.

That's why people sing in the bath – it has the same effect as a concert hall.

Did you know that everything shakes when you speak? If you hold a piece of paper touching your mouth and hum, the paper shakes enough to tickle your lips.

To understand that tickle you need to understand sound; and to do that perhaps you should be present when a note is born.

A NOTE IS BORN

Press a ruler on to a table as shown, with most of it sticking out over the edge. Pluck the end. What happens?

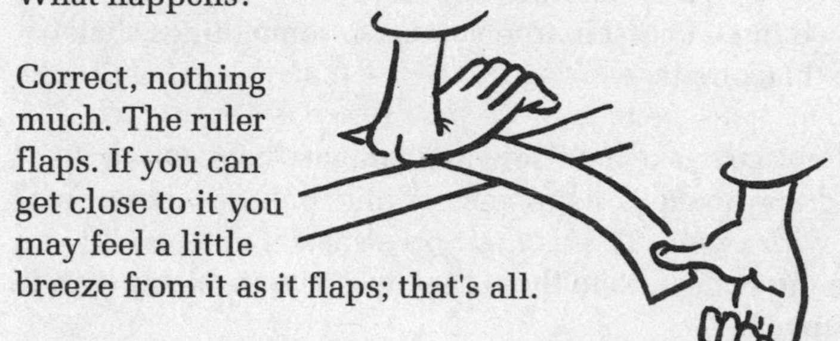

Correct, nothing much. The ruler flaps. If you can get close to it you may feel a little breeze from it as it flaps; that's all.

Move it further on to the table and repeat. The ruler flaps faster this time. Move it up again and repeat . . . and again . . .

At a certain point the vibration will begin to sound like a note. Move the ruler up again . . . it sounds higher. You have witnessed the birth of a note. The tiny breezes find their way into your ear and bounce against the ear drum, which shakes; a series of tiny bones, which pass it on to a sound centre, the **cochlea,** which calls it a note. The faster the note,

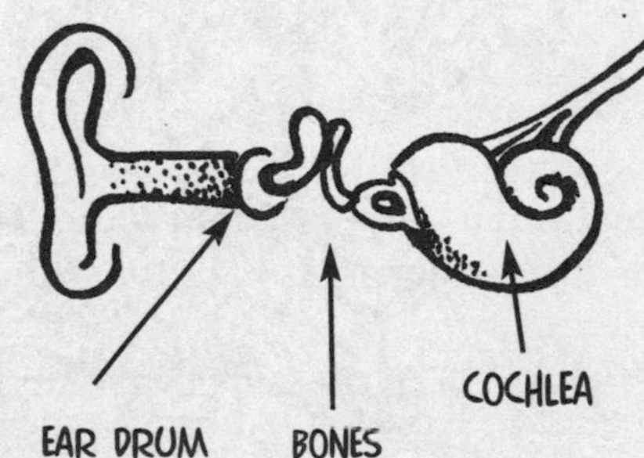

the higher the sound. 'Sound' is the name for the fast stream of tiny breezes, or waves of compressed air.

It isn't only your ear drum that shakes when a sound hits it. So does the wall, the light fitments, the bath, in fact the lot. Only by a tiny amount, of course. But – and here's a funny thing – for each note you utter, some things shake more than others.

Hard objects make a noise when you tap them. Often the sound they make is a shapeless clunk, but they may come back with a note. That is their own special note, and if you make the same note to them they will vibrate along with it, or **resonate.**

Set some wineglasses in a row and fill them with different amounts of water. The water makes them vibrate at different frequencies, so when you tap them with a pencil they each ping with their own note.

GLASS HARP

THE EFFECT

The magician makes a wineglass sing. (Be patient – learning to play the glass harp is rather like learning to whistle; it takes a little practice.)

YOU NEED

- A wineglass half-full of water
- A wet finger

TO PERFORM

Wet one finger and rub it round the rim of a wineglass in a continuous circle. Vary the pressure a little and the wetness a little: when the finger is not so much gliding as rubbing, the glass will begin to sing a clear note, which will fill the room. This is the glass's special note, or **resonant frequency** – the note it will resonate to.

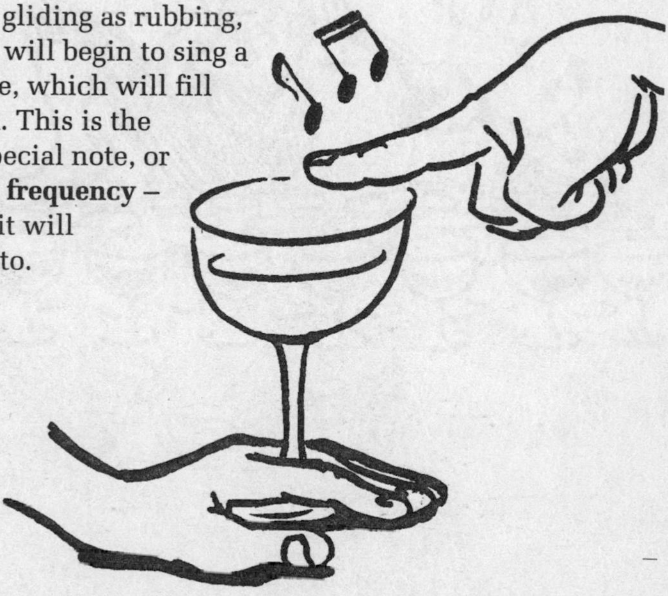

(Have a look at the water's surface as you play – the glass is shaking the meniscus up into a weird pattern.)

You can get different notes if you fill the glass with different amounts of water.

With a little more skill and a lot more glasses you can play tunes that sound like the songs of angels. Mozart wrote several pieces for the glass harp when it was first invented 200 years ago. His glass harp had dozens of glasses in it.

RESONANCE

When a glass is hit by a note of its own personal resonant frequency it vibrates in sympathy. You can prove this by moving two glasses near each other and fill them so that when you rub each one, they sound exactly the same. When you make only one sing, the other will join in. Check it by touching each glass to damp it down.

There are tales of opera singers with voices so pure and so loud that they could break wineglasses just by singing to them. They sang at the resonant frequency of the glass, which shook itself to pieces.

Bathroom Sing-Along

Here's a challenge: can you make the whole bathroom sing? Very often walls, glass doors or baths can made to singalong with you. Put your mouth close to parts of the bathroom and hum a high note quite loudly. Slide down the scale to a low note. If at any time your voice seems to get louder, it is because something near you is resonating at that pitch – adding its voice to yours. Try it in other locations round and about – along school corridors, under bridges, down stair-wells – when you find a good one, the whole town will hear about it.

In the bathroom, the water in your basin has a natural frequency. If you put your hand in it and gently move it from side to side you can find this resonance. It's possible to get the water to slop over either side with very little effort, once you find the rhythm.

When soldiers march across bridges they always break step. If they happened to march at the bridge's resonant frequency, the structure would shake itself to bits.

Microwave ovens transmit electro-magnetic radiation at the resonant frequency of water molecules. All the water in the food starts shaking in harmony, creating heat.

Ancient clocks kept in time by using the resonant frequency of the pendulum. Modern clocks use the resonant frequency of quartz crystals, which vibrate ten thousand times a second.

When you go on a swing in the park, you know that getting a good height is all a matter of timing; playing with the resonant frequency of the swing.

Super Lung

THE EFFECT

This is a test of strength for the lungs. The audience is asked how far they can get a bottle on a string to swing, just by blowing in it. They will only be able to blow it an inch or so; but the magician's Super Lung can get it to swing way up in the air.

YOU NEED

- Bottle - a wine bottle will do
- String - about 1 metre

SECRET PREPARATION

Tie one end of the string to the top of the bottle and the other to a hook, so that the bottle can swing freely.

TO PERFORM

Test everyone's lungs by getting your audience to take it in turns to see how far they can get the bottle to swing, using only the power of puff. (They'll be lucky if they can get it to swing more than a centimetre.) Encourage them all the time, telling them how well they're doing.

Then it's your turn. Give a little puff, and the bottle will swing out a little, then swing back. Give a another little puff as it reaches the back of its travels; it'll swing forward a little further this time.

By blowing at exactly the right moment – in time with the natural swing of the pendulum – you can get the swing up to nearly a metre.

Well done, Super Lung!

WHAT HAPPENED

This is a blown-up, slowed-down version of what a glass does when it sings. In place of the sound waves at 18,000 beats per second we have your puffs at 40 beats per minute, and in place of the glass's resonant frequency, there's the pendulum's natural swing, or **oscillation.**

Harmonious Shampoo Bottles

THE EFFECT

Two shampoo bottles hang side-by-side. The magician sets one swinging, then commands it to stop and the other to start – which they do! Then the magician commands them to change again – which they do again!

SECRET PREPARATION

Tie a length of string loosely between two points, more than 1 metre off the floor, and 1 metre or more apart.

Take two lengths of string about 1 metre long and tie one end of each to the tops of the bottles. Tie the other ends to the high string, about 40cm apart. It's very important that both of these pendulums must be the same length.

TO PERFORM

Tell your audience how all things obey you – except humans, but that's just a matter of time and practice. Shampoo bottles certainly know who is boss.

Set one of the bottles swinging. After a few swings, command the other bottle to start swinging too – it will do so. When it has got under way, command the first bottle to stop. Watch what happens – it will obey you! When the first bottle has stopped swinging, order it to start again, and the second one to stop. Again, they will obey you; and they'll continue to obey for as long as they continue to swing.

WHAT HAPPENED

It should be mentioned that the bottles would do this even if you didn't command them. The rhythm of the first pendulum is carried by the horizontal string to the second one, which picks it up because its natural rhythm is the same. If you hang another bottle in between those two, but on a different length of string, it won't pick up the swing at all because its natural rhythm will be different.

Little Ben

You have noticed how noises can sound different in the bathroom from elsewhere. Compare what you sound like wrapped in a towel to what you sound like with your head in a bucket.

If that seems too ridiculous, do this . . .

THE EFFECT

The magician produces a wire coathanger. When the audience listens to it they hear Big Ben.

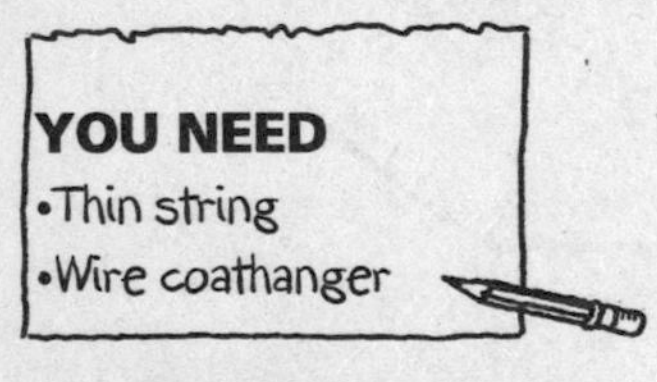

TO PERFORM

Tell the audience that from here, you should be able to hear the chimes of Big Ben. However, failing that, they can always comfort themselves with the sounds of Little Ben.

Produce the wire coathanger and ask for a volunteer.

Get them to wrap the string around each index finger, lean forward and put their fingers in their ears. Hang the hanger in the middle of the string, then tap it with a stick or spoon. You will hear a dull plunk but they will hear Big Ben.

WHAT HAPPENED

The dull plunk of the hanger is a mixture of sounds ranging from low to high notes, all on top of each other. The soft fingers and string absorb the high notes, leaving only the low, Big Ben-ish tones.

When you listen to a recording of your voice, you may be aghast at how thin and reedy you sound, compared to the deep, mature voice that you can hear. But remember, when you listen to your own voice you hear a certain amount of it transmitted to your ears through your head, which is soft. So the high tones are filtered out, and the deep ones get through.

THE MOST FAMOUS BATH OF ALL

2000 years ago, someone had a bath that is still talked about to this day. The King of Syracuse had asked Archemedes, the famous Greek scientist, to discover if his brand new crown was made of pure gold. It felt too light. Had the makers cheated by mixing some cheap metal with the gold?

Archimedes quickly had the answer. Gold was the heaviest metal known to man. Mixing anything with it would result in a lighter crown. So all the king needed to do was melt his crown down into a lump, find a lump of pure gold of the same size and compare their weights. The only snag in this scheme was the bit about melting down that brand new, intricately carved, beautiful crown. This the king would not allow.

Archimedes had to think of a way round the snag.
All day he thought, and even as he filled his bath that night the problem filled his mind. He was so busy thinking that he filled his bath right up to the top, and climbed in, still pondering. Naturally, as he got in, the water poured out all over the floor.

So what did he do? Did he say, 'Oh dear!' or 'Where's the mop?'? Not a bit of it. He looked at the mess on the floor and shouted, 'Fantastic! Yes! Great!'

Then he went dancing down the road, stark naked,

shouting, 'Eureka! Eureka! ('I've got it!' in Greek).
What Archimedes had realised was that the amount of water that poured out of the bath was exactly equal to the amount of him that got into it. So if he dropped the king's crown in the bath, the spillage would be equal to the amount of metal in the crown.

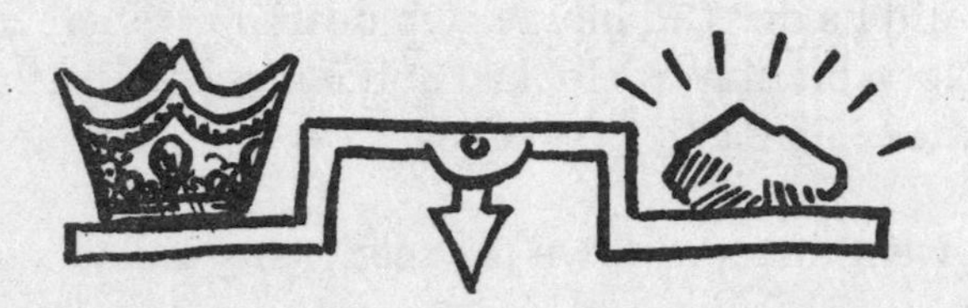

All he had to do was find a lump of gold the same weight as the crown ..

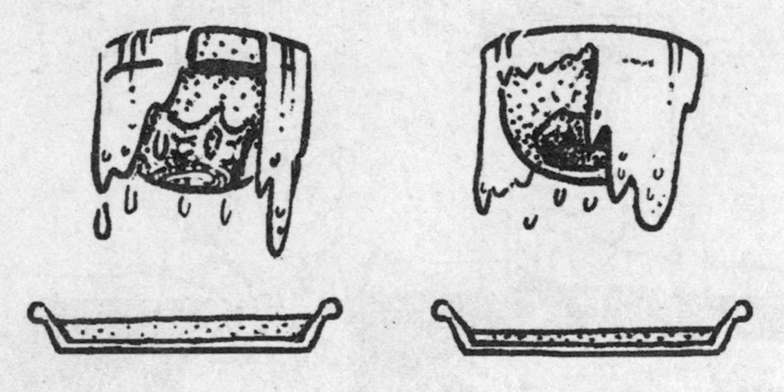

and see if it spilled the same amount of water.

If it did, then they were both made of gold. If it spilled less, the crown was a fake.

Next time anybody complains about you spilling water on the floor, tell them you are practising for when the King of Syracuse asks your advice about his brand new throne.

Meanwhile, enjoy your bath!

LIST OF MAGIC TRICKS